Emelda Orakwue, Christine Murangwa

Constructed Wetland for municipal sewage wastewater treatment in Middle East/North Africa [MENA] (Iran)

GRIN Verlag

Bibliografische Information der Deutschen Nationalbibliothek:

Die Deutsche Bibliothek verzeichnet diese Publikation in der Deutschen National-
bibliografie; detaillierte bibliografische Daten sind im Internet über http://dnb.d-
nb.de/ abrufbar.

Impressum:

Copyright © 2013 GRIN Verlag GmbH
Druck und Bindung: Books on Demand GmbH, Norderstedt Germany
ISBN: 978-3-656-44222-6

Dieses Buch bei GRIN:

http://www.grin.com/de/e-book/215591/constructed-wetland-for-municipal-sewage-
wastewater-treatment-in-middle

Wetlands for Water Quality

A design proposal by

Emelda Orakwue

Christine Murangwa

Constructed Wetland for municipal sewage wastewater treatment in Middle East/North Africa [MENA] (Iran)

31/03/2013

Contents

1. **Introduction**

1.1 Case study description

The Middle East and North Africa (MENA region) is located between longitudes 13^0 W and 60^0 E between latitudes 15^0 N and 40^0 N covering a surface area of about 11.1 million square kilometers or about 8% of the area of the world (Fig. 1). MENA is the home for about 300 million people, or about 5% of the world's population, with an average annual population growth rate of 1.7% (World Bank, 2005). The 21 countries in the MENA region have many similarities of arid conditions (85% of the area is desert). Water resources are adversely affected as a result of changes in climate, population, economic development and environmental considerations. The MENA region can be considered as the most water-scarce region of the world (Droogers et al., 2012).

Iran (one of the countries in the Middle East located in southwest Asia) was used as the case study because certain parameters (precipitation, evapotranspiration, temperature and population equivalent) used were derived from this country. Iran has a complete desert climate that is very hot in the summer and relatively cold in the winter with scarce rainfall of 200 mm/year and high evapotranspiration (Droogers et al., 2012). Water crisis has appeared in Iran as a serious problem. There are reasons for that: (1) Lack of proper water management and (2) Occurrence of drought (Motiee et al., 2000)

Consequent to severe water scarcity, there is need for treatment of the generated wastewater for reuse. Constructed Wetland is one of the secondary treatments for treating primary effluent. This was designed to treat municipal sewage wastewater for one thousand persons in Iran.

1.2 Area location

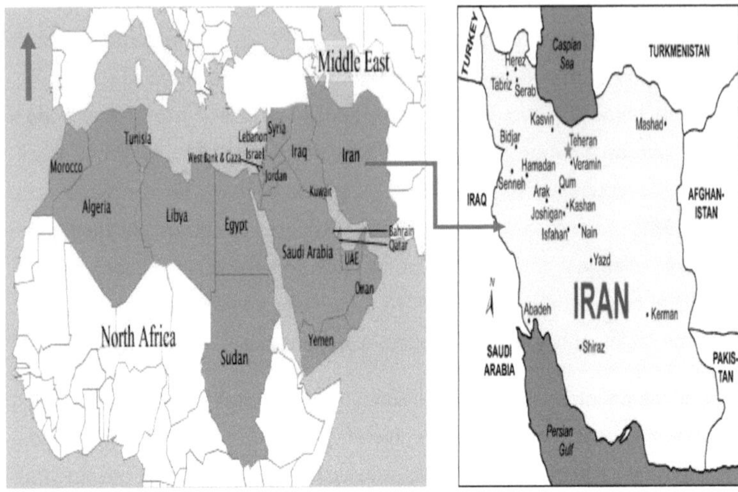

Figure 1.

Sources: *http://forum.nationstates.net/viewtopic.php?f=20&t=211522&start=125 for MENA map and http://www.jacobsenrugs.com/iran.htm for Iran map*

1.3 MENA - Iran Condition

Women fetching water from a shallow depth muddy flowing stream.

Figure 2.

A river already dried due to high evaporation

Figure 3.

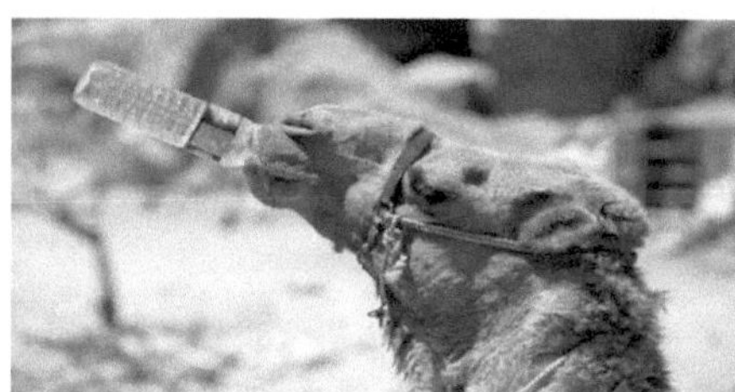

Camel drinking bottle water because of scarcity of natural water body

Figure 4.

Woman fetching water from almost dried water pond

Figure 5.

1.4   Objectives

Main objective is to design an economically feasible and sustainable constructed wetland for wastewater treatment in Iran (MENA region).

This could be achieved by

a. The reuse of the treated wastewater effluent for irrigation (vegetable production)
b. The reuse of the treated wastewater for fish pond

2. Design

Figure 6.

Source: *http://www.kibbutzlotan.com/community/ecology/wetlands.html#.UVcftldiLXo*

Constructed wetlands are man-made systems, designed and constructed to treat wastewater using the natural processes typical of natural wetlands. These natural processes are an interaction and combination of wetland plants (reeds), substrates (sand, gravels etc) and microbial organisms. Knowledge of these natural processes is used to control the operation and efficiency of the constructed wetlands. Constructed wetlands can be classified according to the water flow regime: surface flow systems and subsurface flow systems. Constructed wetlands with subsurface flow may be classified according to the direction of flow into horizontal subsurface flow (HSSF) and vertical subsurface flow (VSSF).

2.1 Basics for CW design

Factors considered before the design:

a. Wastewater type: municipal sewage wastewater is to be treated.
b. Wastewater quantity
c. Soil type: Sand, gravel and coarse sand with iron grits were used for the design.
d. Flows (water balance)

   $Q_o = Q_i + (P\text{-}ET) \times Area$

   Q_o = Wastewater outflow

   Q_i = Wastewater inflow

   P = Precipitation

   ET = Evapotranspiration

e. Concentrations (Influent and Effluent)
f. Constructed wetland type: Because the effluent of the treated wastewater will be used for irrigation and fish pond, horizontal subsurface flow system was selected. This selection type was also made to avoid breed for mosquito nuisance. Additionally to avoid evaporation and also this system is easy to operate and maintain.
g. Landscape : The area is mainly of rugged, mountainous rims surrounding high interior basins
h. Budget:

2.2 Design flowchart

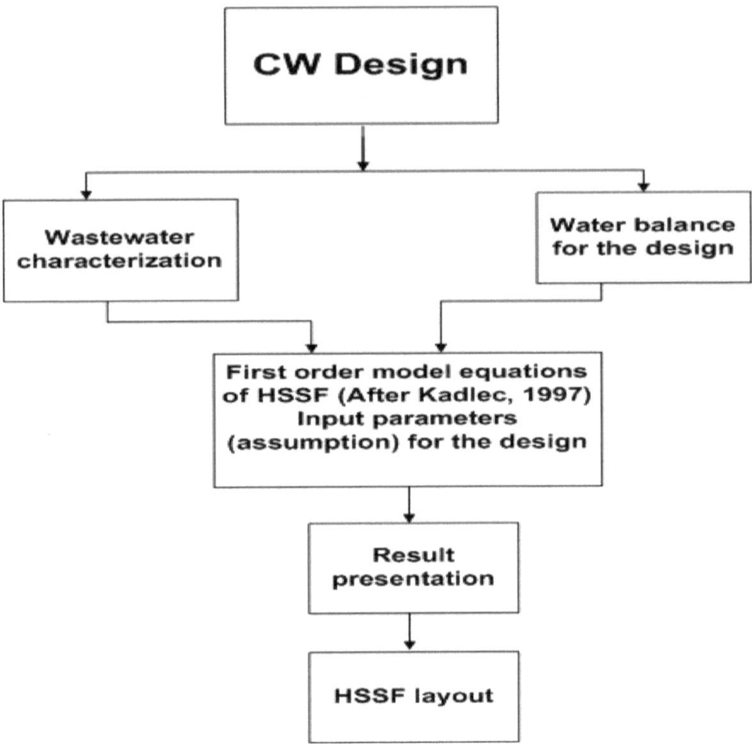

This flowchart explains the way by which the design was carried out. The wastewater to be treated was firstly characterized and water balance was calculated. Secondly, the first order model equations of HSSF (Kadlec, 1997) were used to input assumed parameters for the design. Then when these parameters were calculated, their results were shown and lastly the layout of the designed HSSF system.

Table 1: Wastewater characterization

Parameter	Units	C_i Influent (Primary effluent)	C_o Effluent
BOD_5	mg/L	150	30
SS (Suspended solids)	mg/L	120	30
T-N (Total Nitrogen)	mg/L	50	40
T-P (Total phosphorus)	mg/L	9	8
FC (Faecal Coliforms)	cfu/100ml	10^7	10^2

Biological oxygen demand for five days (BOD_5), suspended solids, total nitrogen, total phosphorus and faecal coliforms were the five parameters characterized in the municipal sewage wastewater. Their influent and effluent concentrations were also shown.

Water Balance

Table 2: Performance-first order for HSSF (Kadlec, 1997) Assumption

	BOD	SS	T-N	T-P	FC
$K_{20}^{0}C$ (m/year)	180	1000	27	12	95
C^* (mg/L	3.5 + 0.053Ci=11.43	7.8 +0.063Ci = 17.25	1.5	0.02	10
θ	1.00	1.00	1.05	1.00	1.00

$$Q_o = Q_i + (P\text{-}ET) \times Area$$

$$P = 200 \text{ mm/year}$$

$$ET = 255 \text{ mm/year}$$

$$Q_i = 76 \text{ m}^3/\text{day}$$

Before Area could be calculated K_T and q values have to be calculated first.

$$\mathbf{K_T = K_{20} \times \theta^{(T\text{-}20)}}$$

$$\mathbf{q = - K_T/\ln [C_e - C^*/C_i - C^*]}$$

Where K_T is first order rate constant at temperature

T: temperature for Iran [19.9^0C] (Mozaheb et al., 2010)

q: Hydraulic loading rate

PE (Population Equivalent) = 1000 persons

Average water consumption = 95L/day/PE

Q_i = 95L/day/PE x 1000PE x 80% = 76 m^3/day

Water quantity to be treated is **76 m^3/day**

$$\mathbf{A = Q/q}$$

Hydraulic retention time: HRT = V/ Q

Because Evapotransipiration is higher than precipitation in the study area (ET> P), these values were neglected and assumed to be zero.

Q_o = 76 m^3/day

Q_i = 76 m^3/day

Table 3: Result presentation

Parameters (mg/L)	K_T(m/d)	q (m/d)	Area (m^2)	Load (kg/d)	Load cfu/100ml
BOD	180	0.2453	310	11.4	
SS	1000	1.3129	57.9	9.12	
T-N	26.87	0.3188	238.4	3.8	
T-P	12	0.2785	272.89	0.684	
FC	95	0.0224	**3400**		7.6×10^{8}

The largest area which is that of faecal coliform was selected for the design; this means that the area will be sufficient enough for the removal of all the above mentioned parameters.

Table 3: Result presentation

Calculated values		
Actual Volume (m^3) = 1700	Actual vol. + SF = 1870	Safety factor = 10%
HRT (Days)	25	

Table 4: Result presentation

Assumed values		
Length (L)	142.8m	Let L = 6W solved using
Width (W)	23.8m	LxW=A
Depth (D)	0.5m	

Figure 7: Horizontal subsurface flow system layout

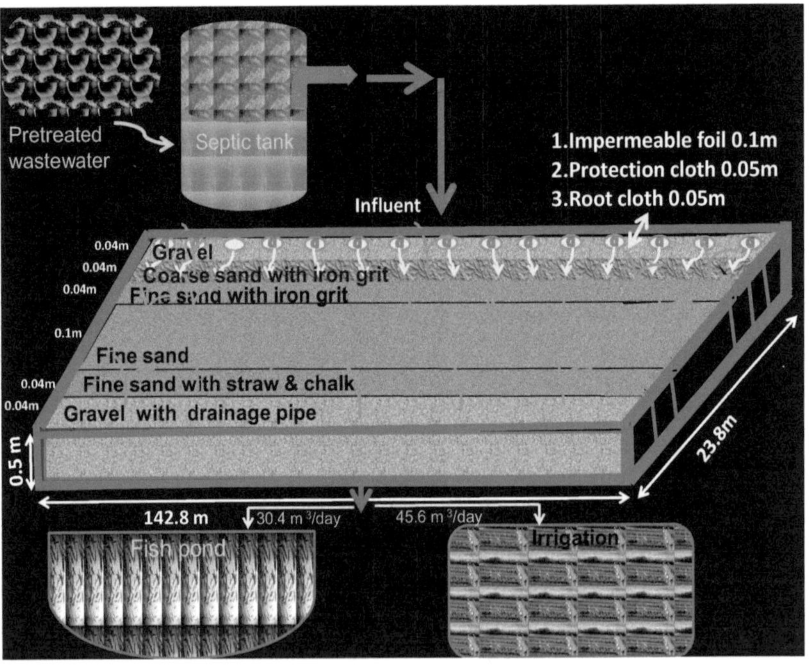

The raw municipal sewage wastewater will first undergo pretreatment by removal of large particles, oil and grease so as to avoid damage to the primary treatment equipment. The pretreated wastewater will be channeled to a septic tank where sedimentation for primary treatment will take place. Then the primary effluent is finally channeled through the length side of the design using fifteen pipes so as to minimize dead zones encounter and also to avoid clogging of the wetland. The principal reason for various channels also is to evenly distribute the wastewater in the constructed wetland.

The layout is rectangular in shape with length, width and depth of 142.8 m and 23.8 m and 0.5 m respectively. The substrates (sand, gravel and coarse sand with iron grits) and their various ranges were used for the design. Phragmiteses (reeds) were planted on the HSSF CW because these plants type can survive severe weather conditions and their roots can also extend deeper in the ground thereby having firm support, and also reed plants help in oxygen transfer. Straw and limestone are used for phosphate removal and carbon source for microorganisms for nitrification and denitrification.

The organic components in the wastewater are biodegraded as they pass through the substrates. Also plants take up nutrients (Nitrogen and phosphorus) during the wastewater passage. About 60% (45.6 m^3) and 40% (30.4 m^3) of the effluent are used for irrigation and fish pond respectively.

2.3 Investment cost

Table 5: Investment cost on items used for the design construction

Item Description	Quantity	Price (€)	Total Cost (€)
Land	3400m^2	0.5/m^2	3400
Digging (CW)	-	-	2500
Plastic Covers	3 kits	500	1500
Plants	1550	0.5	775
Sand soil	300m^3	1/m^3	300

Septic Tank	1	1000	1000
Pipes	30	25	750
Irrigation operating system	-	-	1000
Fish pond	1	1000	1000
Labor	-	-	1500
Accessories	-	-	1000
Protecting trees	150	2	300
pH/DO/EC meter	2	350	700
Total Cost			**15,725**

The investment cost is highly dependent on land and substrate purchase. From this investment cost list it can be seen that land and substrate has the highest purchased price.

3. Management plan

3.1 Operation and Maintenance

Table 5: Categories of operation and maintenance activities (UN-HABITAT, 2008)

Daily activities	Monthly activities	Annually activities
-Monitoring of flow -Water level -Water quality -Checking inlet structure -Checking outlet structure	-Vegetation management -Removal of suspended solids -Control structure -On-going operational	-Vegetation harvesting -Checking and cleaning of distribution system -extraction of sludge

	records	

For an effective and continuous working of a constructed wetland, some operation and maintenance activities could be done daily, monthly and annually as shown above.

Table 6: Operation and Maintenance cost

Items	Costs (€)
Inlet and outlet monitoring (general maintenance)	1500
Vegetation management	600
Desludging	1200
Total	**3300**

3.2 Benefits and Revenue

Benefits

The constructed wetland provides us with various provisional (fish, vegetable and reeds [local roofs]), regulating (water purification) and supporting services (habitat for species).

Table 7: Revenue

Product	Amount	Price (€)	Revenue (€/y)
Fish	10 kg	2	10*30*12*2=7200
Irrigation water	35m^3	15.6	35*30*12*15.6=196560
Vegetable	10 kg	7.8	10*30*12*7.8=28080
Total			**231,840**
Net benefit:231840-(15725+3300)= **212,815(€/y)**			

4. Conclusion & Recommendation

Conclusion

Based on the net benefit from the HSSF CW it can be seen that the benefits generated is far higher than the investment cost, thus making the CW economical and sustainable.

Also effluent is reused for fish pond & irrigation by this means improving the life of MENA region people.

Recommendation

We strongly recommend that this project should be implemented in MENA region especially in Iran. Also that land should be made available for the construction in order to minimize the investment cost.

References

Droogers, P., Immerzeel, W. W., Terink, W., Hoogeveen, J., Bierkens, M. F. P., Van Beek, L. P. H., Debele, B., 2012. Water resources trends in Middle East and North Africa towards 2050. Hydrology and Earth System Sciences, 16, 3101-3114.

Handouts and lecture notes on Constructed Wetlands for Water Quality, 2013. Module 6.

Kadlec, R. H., 1997. Deterministic and stochastic aspects of constructed wetland performance and design. Water Science and Technology, 35, 5, 149-156.

Motiee, H., Manouchehri, G. H., Tabatabai, M. R. M., 2000. Water Crisis in Iran: Codification and Strategies in Urban Water. Water and Wastewater Eng. Dept., P. O. Box 16765-1719, Tehran.

Mozaheb, S. A., Ghaneian, M. T., Ghanizadeh, G. H., Fallahzadeh, M., 2010. Evaluation of the Stabilization Ponds Performance for Municipal Wastewater Treatment in Yazd - Iran. Middle-East Journal of Scientific Research, 6, 76-82.

UN-HABITAT, 2008. Constucted Wetlands Manual. UN - HABITAT Water for Asian Cities Programme Nepal, Kathmandu. United Nations Human Settlements Programme, 978-92-1-131963-7.

World Bank: Middle East and North Africa Data Profile, Based on World Development Indicators database, Washington DC, USA, August, 2005.